The Nature of Matter

by Christine Caputo

Table of Contents

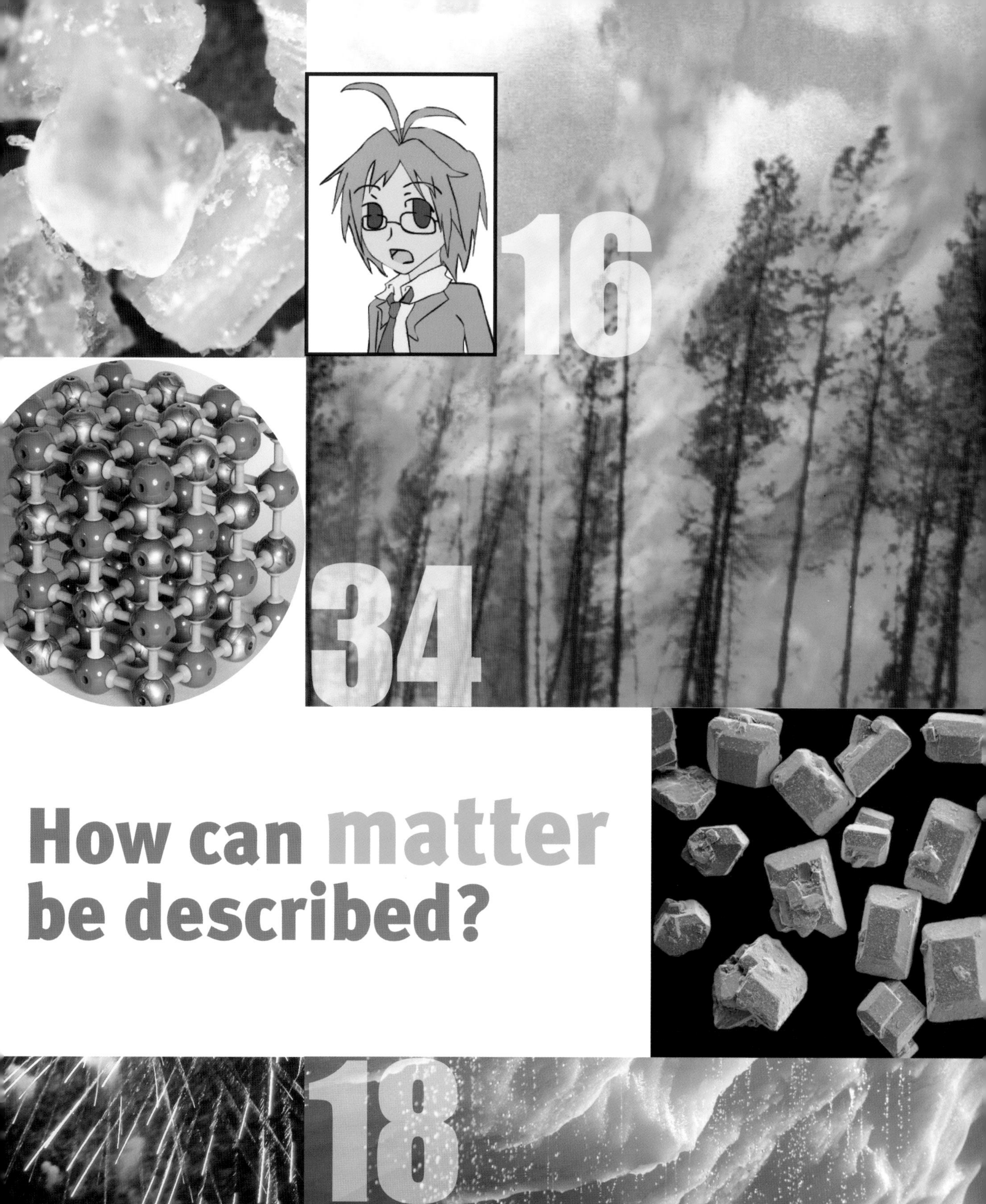

How can matter be described?

INTRODUCTION

Last Flight of the Hindenburg

The year was 1937. The *Hindenburg* left Germany. The airship was going to New York City. The airship was about to land in New Jersey. Just then, the airship caught on fire! People were scared. People below ran for safety. Seconds later, the burning ship was on the ground. Thirty-six people died. What happened? What caused the airship to burn?

Now people fly on jet planes. Some of the first airships were called dirigibles (DEER-ih-jih-bulz). These airships were like large balloons. People thought these huge ships were the modern way to travel. These ships had steel structures. These ships were full of a gas called hydrogen. This gas made the ships float in the air.

Within just thirty-seven seconds, the ship's burning wreckage lay on the ground.

In 1935, the Zeppelin Company designed the *Hindenburg*. The largest aircraft ever flown, it was 804 feet long, carried more than 1,000 passengers, and could lift 112 tons.

In the early 1900s, airplanes could travel only short distances before stopping for fuel and could carry very little weight. Airships were considered to be the way of the future.

▲ On May 6, 1937, the age of dirigibles came to a tragic end.

One problem with hydrogen is that it can catch fire quickly. The paint on the ship also burned easily. May 6 was a stormy day. Electricity from the storm set fire to the paint. The fire from the paint lit the gas inside the airship.

What if the airship had been filled with a gas that does not burn? What if the airship had a different type of paint? Maybe the airship would not have burned.

Today, the *Hindenburg* gives us a tragic lesson in the nature of matter. It is important to understand how matter behaves. Knowing how matter behaves can keep people safe.

CHAPTER 1

Types of Matter

What makes matter an element, a compound, or a mixture?

Matter (MA-ter) is all around us. The objects you touch are matter. The foods you eat are matter. The air you breathe is matter. Even the people around you are matter. Matter has **mass** (MAS) and **volume** (VAHL-yoom). Mass is the amount of matter in an object. Volume is the amount of space an object takes up.

Roller coasters, waterslides, and helium-filled balloons are all examples of matter.

Some matter is shiny. Some matter is dull. Some matter can bend. Some matter can break. Some matter is easy to see. Some matter we cannot see. What makes one type of matter different from another? The answer depends on what makes up the matter.

Elements

The simplest type of matter is an **element** (EH-leh-ment). An element cannot be broken down into something else. We find some elements in nature. We make some elements, too. Oxygen and carbon are two elements we find in nature.

Essential Vocabulary

atom page 10
chemical formula page 12
chemical symbol page 8
compound page 10
element page 7
mass page 6
matter page 6
mixture page 13
molecule page 11
pure substance page 13
volume page 6

Chemical Names and Symbols

Each element has a name. Some of these names come from Latin or Greek words. Other elements are named after famous people and places. One element is named after Albert Einstein. Americium is named after America. Can you guess what Neptunium is named after?

Every element has a symbol. Symbols help people everywhere identify elements.

A **chemical symbol** (KEH-mih-kul SIM-bul) is an abbreviation. A chemical symbol has one or two letters. The first letter is always uppercase. The second letter is always lowercase. You can see the chemical symbols on the chart below.

Al stands for aluminum. B stands for boron. Other symbols can be hard to remember. Hg stands for mercury. The Greek word for mercury is *hydrargyrum*.

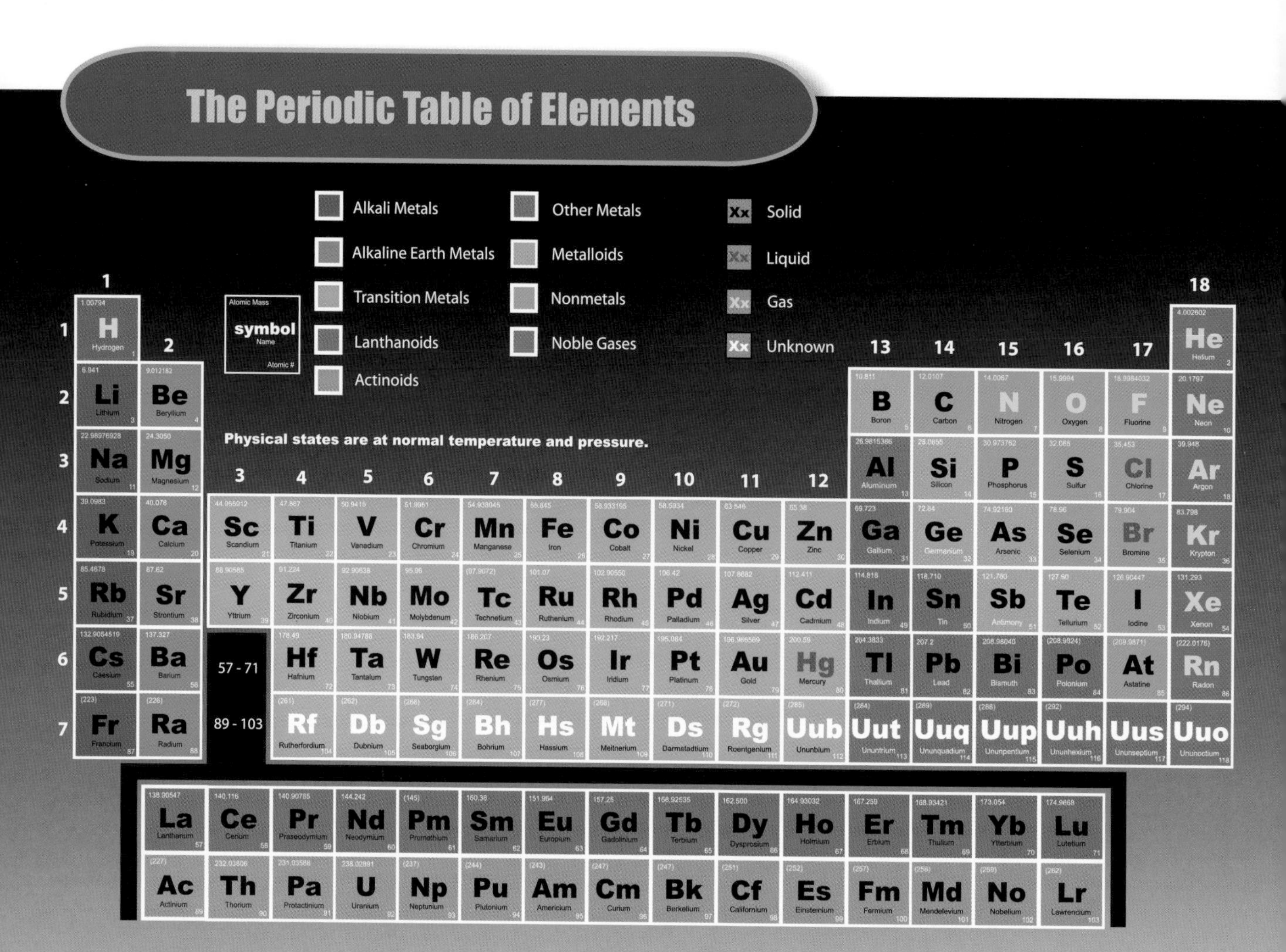

▲ Each element can be described by its name and its chemical symbol.

Science to Science

EARTH SCIENCE AND LIFE SCIENCE

Just five elements make up more than ninety percent of Earth's crust. These elements are oxygen, silicon, aluminum, iron, and calcium. Just three elements make up over ninety percent of the human body. They are oxygen, carbon, and hydrogen.

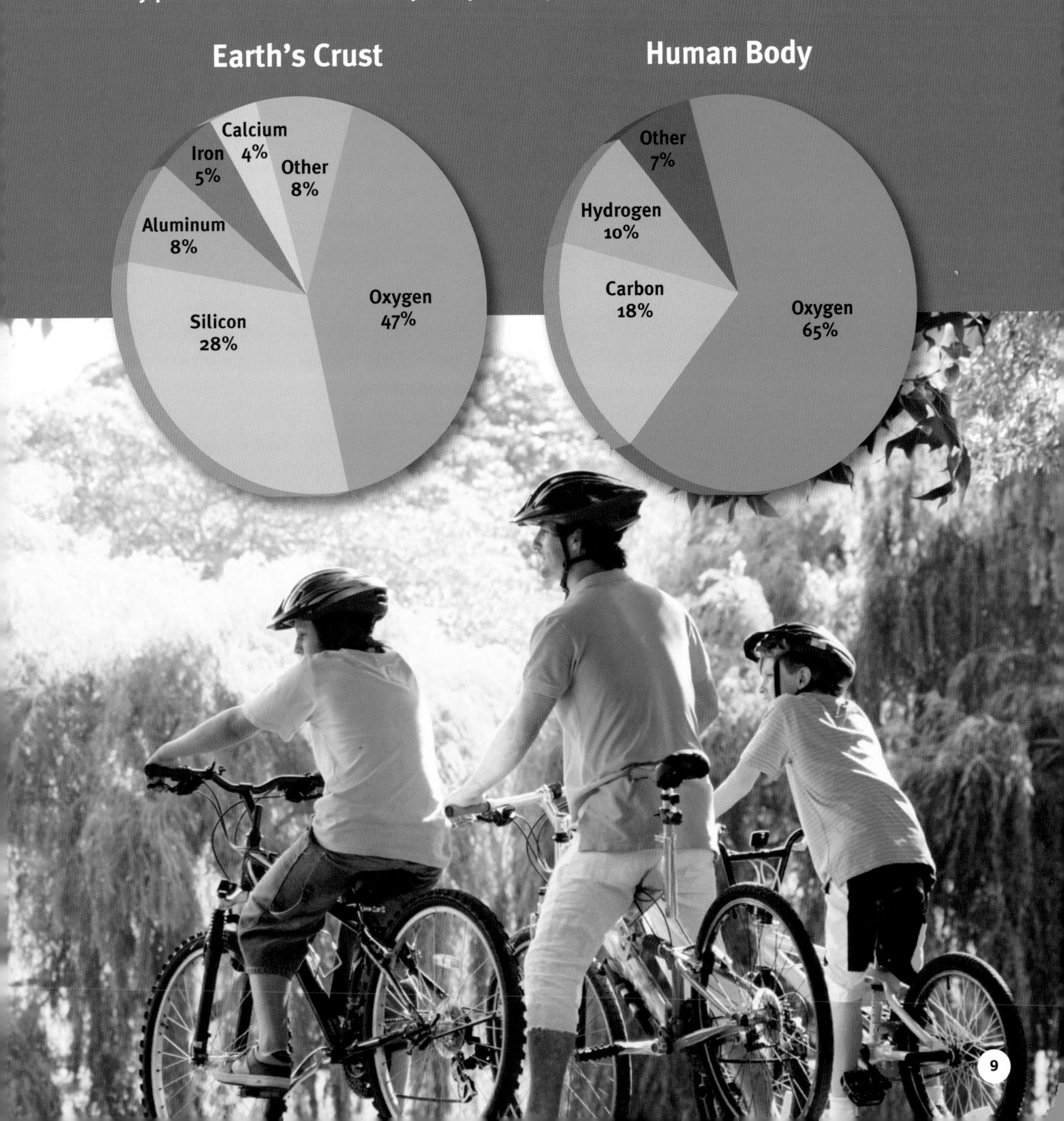

Atoms

The most basic part of an element is an **atom** (A-tum). An atom is the smallest part of an element. People call atoms the building blocks of matter. Atoms are tiny. More than one million atoms can fit on the head of a pin! Each element is made up of one type of atom. Gold is made up of gold atoms. Copper wire is made up of copper atoms. Neon gas is made up of neon atoms.

All atoms have a similar structure. An atom is made up of small particles. These particles are called protons, neutrons, and electrons. Protons and neutrons are in the nucleus. The nucleus is the center of an atom. Electrons move around the nucleus. Atoms of different elements have different numbers of particles. The different number of particles make elements special.

▲ An atom of carbon has:

- six protons
- six neutrons
- six electrons

Compounds

When atoms combine, forces of attraction hold them together. These forces are called chemical bonds. When the atoms of two or more elements combine by chemical bonds, a **compound** (KAHM-pownd) forms. Ammonia is a compound. Ammonia has three hydrogen atoms joined to one nitrogen atom. Other compounds are sugar, salt, chalk, and water.

Every water molecule on Earth is made up of two hydrogen (H) atoms and one oxygen (O) atom. Similarly, every molecule of nitrogen is made up of two nitrogen (N) atoms. Nitrogen gas makes up about seventy-eight percent of the air you breathe.

The Root of the Meaning:
THE WORD
COMPOUND
comes from the Latin *comp nere,* which means "to put together."

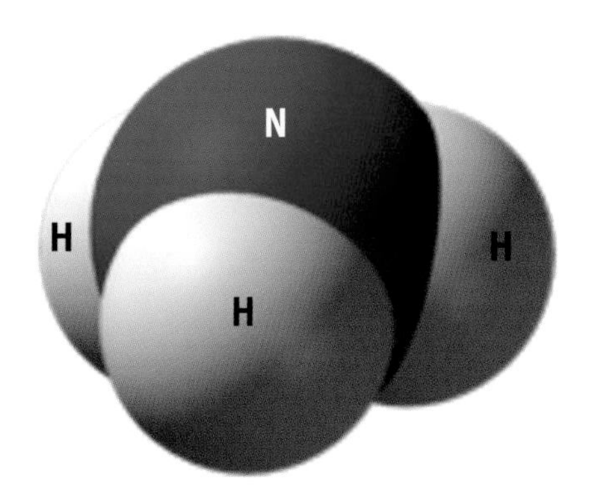

ammonia molecule (NH_3)

▲ The white spheres represent hydrogen (H) atoms. The blue sphere in the center represents a nitrogen (N) atom. Together they represent the compound ammonia (NH_3). Ammonia has many important uses, including the manufacture of fertilizers used to grow crops.

Molecules

Many compounds are **molecules** (MAH-leh-kyoolz). A molecule is the smallest unit of an element or compound that has all of the properties of that element or compound. Water is a molecular compound. Each water molecule has two hydrogen atoms bonded to one oxygen atom. Some molecules are made up of only one type of atom. A nitrogen molecule forms when two nitrogen atoms bond together. Because a nitrogen molecule consists of only one type of element, it is not a compound.

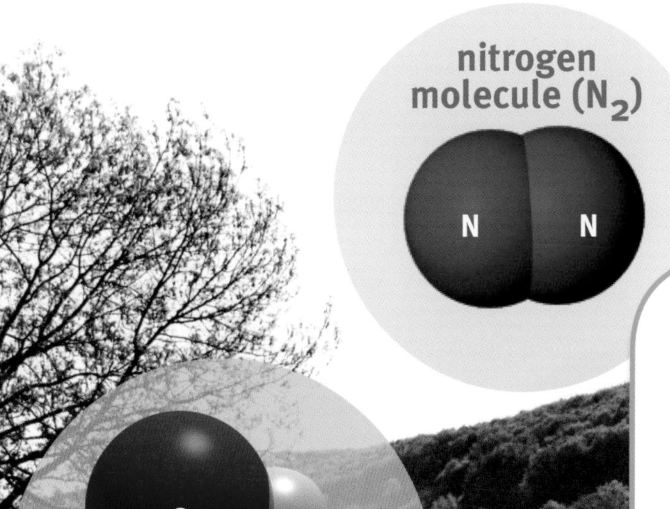

nitrogen molecule (N_2)

water molecule (H_2O)

CHECKPOINT

Visualize It

Compounds are commonly represented by models. These models are often made up of colored spheres, or balls. Each sphere represents an atom. In this book, oxygen atoms are red and hydrogen atoms are white. Work with a partner to develop your own model for water using whatever materials you have available, even jellied candies or marshmallows. Explain what each object in your model represents.

Chemical Formulas

A compound has a **chemical formula** (KEH-mih-kul FOR-myuh-luh). A chemical formula shows us the elements in a compound. A chemical formula also shows how many atoms of each element are in a compound.

We use chemical symbols when we write a chemical formula. A molecule of carbon dioxide has one carbon atom. This molecule also has two oxygen atoms. The symbol for carbon is C. The symbol for oxygen is O. The formula starts with the symbols for each element. Then we add a subscript to each symbol when the number of atoms is greater than one. A subscript is a small number written to the right of a symbol. If there is only one atom, you do not write a subscript. The subscript is understood to be one. So, the chemical formula for carbon dioxide is CO_2.

Science and Math

Ratios

A ratio compares two numbers. You can write a ratio as a:b, a to b, or a/b. In a compound, you can find the ratio of combining atoms by looking at the subscripts in a chemical formula.

What is the ratio of carbon (C) to hydrogen (H) in methane, CH_4?

Answer is 1:4.

▲ Every molecule of carbon dioxide has one carbon (C) atom and two oxygen (O) atoms. The chemical formula for carbon dioxide is CO_2.

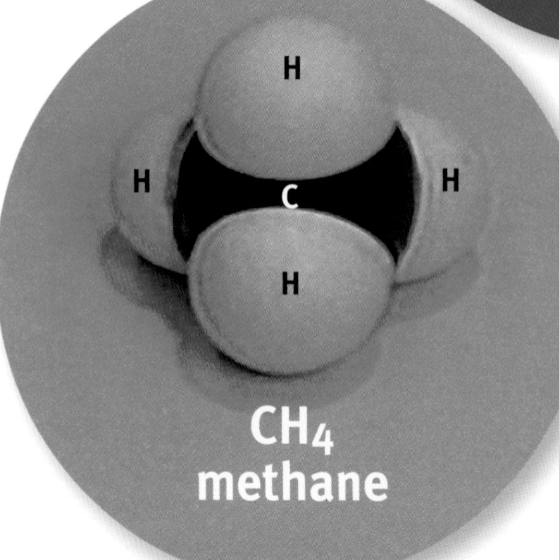

▶ Every molecule of methane has one carbon (C) atom and four hydrogen (H) atoms. The chemical formula for methane is CH_4.

Mixtures

Elements and compounds are called **pure substances** (PYER SUB-stans-es). All molecules in a pure substance have the same properties. All samples of a certain element have the same properties. This is the same for compounds, too. For example, all samples of pure gold have the same properties. All samples of pure water have the same properties. Most of the matter you see each day is not made of pure substances. Most matter is made up of mixtures. A **mixture** (MIKS-cher) is a blend of two or more substances. The parts of a mixture are not combined in a special way. Each substance in a mixture keeps its own properties.

Ocean water is a mixture. Ocean water has water, salts, and other things. Water from the Atlantic Ocean will not be the same as water from the Pacific Ocean. One sample might have more salt than the other. Even samples from the same ocean may differ.

Almost any example of matter you can think of is a mixture. Even the purest of materials contains small amounts of other substances. The sand on the beach is a mixture, as are ocean water and even the air above it.

Heterogeneous Mixtures

We can see the different parts of a heterogeneous (heh-tuh-ruh-JEE-nee-us) mixture. The parts of a heterogeneous mixture are not evenly mixed. Beach sand, concrete, and fruit salad are heterogeneous mixtures.

Homogeneous Mixtures

We cannot see the different parts of a homogeneous (hoh-muh-JEE-nee-us) mixture. Ocean water is a homogeneous mixture. Air is a homogeneous mixture. Air is a mixture of different elements.

A solution (suh-LOO-shun) is a type of homogeneous mixture. One substance is dissolved into another in a solution. The solvent (SAHL-vent) dissolves the solute (SAHL-yoot). Simple syrup is a type of mixture used in cooking. This syrup is a solution of water and sugar. Water is the solvent and sugar is the solute.

Science and Technology

Desalination

In some places where fresh water is scarce, people use desalination (dee-sa-lih-NAY-shun) to remove dissolved salts from seawater. One method involves pumping seawater through thin layers of specially made materials called membranes. The dissolved salts get caught on the layers while the fresh water passes through. Another method involves heating the water until it evaporates and the dissolved materials are left behind.

Unfortunately, the costs of desalination are high. Researchers are working to perfect methods that use less energy in hopes of making this technology more widely available.

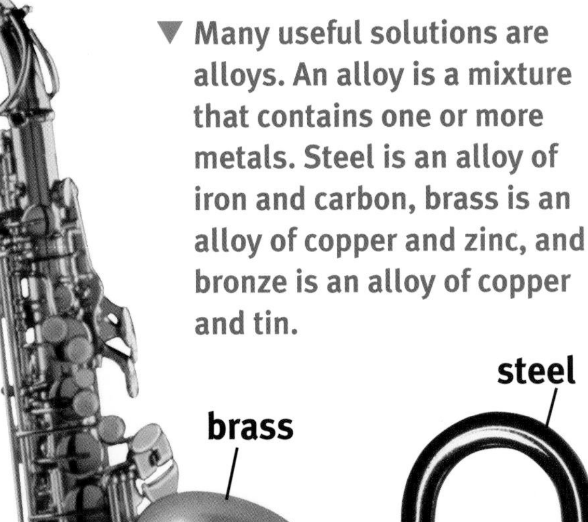

▼ Many useful solutions are alloys. An alloy is a mixture that contains one or more metals. Steel is an alloy of iron and carbon, brass is an alloy of copper and zinc, and bronze is an alloy of copper and tin.

Summing Up

- Matter is anything that has mass and volume.
- The characteristics of matter depend on the elements from which it is formed.
- An element is a pure substance that cannot be broken down into simpler substances. An element is made up of only one type of atom. Each element is represented by a chemical symbol.
- Atoms of elements can bond together to form compounds and molecules, which are represented by chemical formulas.
- Two or more substances blended together physically make up a mixture.
- The different parts of a heterogeneous mixture can be seen because they are unevenly mixed. The parts of a homogeneous mixture are blended so well that they cannot be seen. One type of homogeneous mixture is a solution.

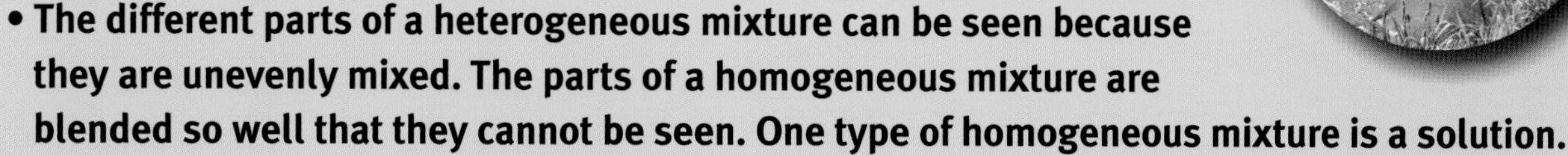

Putting It All Together

Choose one of the research activities below. Work independently, in pairs, or in a small group. Share your responses with the class. Listen to other groups present their responses.

1. Write a paragraph that describes atoms, molecules, elements, and compounds in your own words. Discuss how each type of matter is different and how they are related to one another.

2. In a few paragraphs, explain how a heterogeneous mixture is different from a homogeneous mixture. Create a chart, collage, or drawing that shows different examples of each type of mixture.

3. Research the parts of soil and how they can vary. Explain why soil is a mixture and find out how different mixtures of soil are related to plant growth. Present your findings in a report.

HUMAN NATURE vs. THE NATURE OF MATTER

CARTOONIST'S NOTEBOOK
ILLUSTRATED BY KYLEE SOLARI

Matter exists all around us.

Find examples of the three major states of matter in your room.

What properties can you use to describe each example?

CHAPTER 2

States of Matter

How is the state of matter related to the motion and arrangement of its particles?

Look at this picture. Can you find three forms of water? We see liquid water in the brook. The snow is frozen water. The air has water, too. Matter can be found in different physical states. The three main states of matter are **solids** (SAH-lidz), **liquids** (LIH-kwidz), and **gases** (GAS-ez). **Plasma** (PLAZ-muh) is a state of matter, too.

Essential Vocabulary

energy page 26
gas page 19
liquid page 19
plasma page 19
solid page 19

In this wintry scene, water exists in three different states—solid, liquid, and gas.

CHAPTER 2

Solids

Solid matter has a definite shape. Solid matter has a definite volume. The particles of a solid are packed together. The particles are held in fixed positions. Attractive forces hold the particles still. These forces stop particles from moving past one another.

▲ This solid ice sculpture has a definite shape and volume. If the sculpture is moved, neither its shape nor its size changes. The molecules are packed so tightly together that they barely have room to vibrate.

Crystalline Solids

Some solids form in regular patterns. These patterns are called crystals. Solids made of crystals are called crystalline (KRIS-tuh-lin) solids. Salt, sugar, and quartz are crystalline solids.

The mineral quartz ▶ (SiO_2) is a crystalline solid. Paraffin is an amorphous solid. Their structures give these solids different characteristics.

Amorphous Solids

Some solids do not form in regular patterns. These solids are called amorphous (uh-MOR-fus) solids. Plastics, wax, and rubber are amorphous solids. Mayonnaise, ketchup, and cotton candy are amorphous solids, too.

The Root of the Meaning:

THE WORD **AMORPHOUS** comes from the Greek *amorphos,* which means "shapeless."

Science and Technology

Aerogels

The lightest known solids are called aerogels. Their densities are over 500 times less than the density of water. Scientists began making aerogels in 1931. Back then, they were very hard to work with. But today, scientists have found ways to test aerogels for a variety of uses, including the foam in refrigerators, insulation for windows, and devices used to catch particles in outer space.

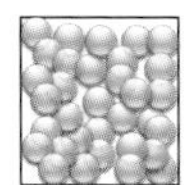 Liquids

Liquid matter has a definite volume. Liquid matter does not have a definite shape. A liquid takes the shape of its container.

The particles of a liquid are not packed as closely as the particles of a solid. Liquid particles can move past one another. Liquid can flow.

▼ Particles in a liquid are able to flow past one another. Liquid particles have greater motion than those in a solid, but they are still packed closely together.

✔ CHECKPOINT

Visualize It

Gather several marbles or small balls. With a partner, use the objects to model solids and liquids. Represent both the arrangement and motion of the particles of matter in each state. Present each model to your class. Challenge your classmates to identify which state of matter you are modeling. Ask your classmates to suggest any improvements you might make to your models.

▲ All three containers are holding the same volume of liquid. However, the shape of the liquid depends on the shape of the container.

CHAPTER 2

Surface Tension

Each particle of a liquid attracts all of the particles around it. The particles on the top of a liquid do not have other liquid particles above them. The other particles of the liquid pull the surface down. This makes the particles on the surface pull together more tightly. This creates surface tension. Surface tension is a force across the surface of the liquid. This force causes a liquid to make drops.

Surface tension causes the surface of a liquid to act like a thin skin. This allows insects to walk across water. Surface tension allows leaves to float on water.

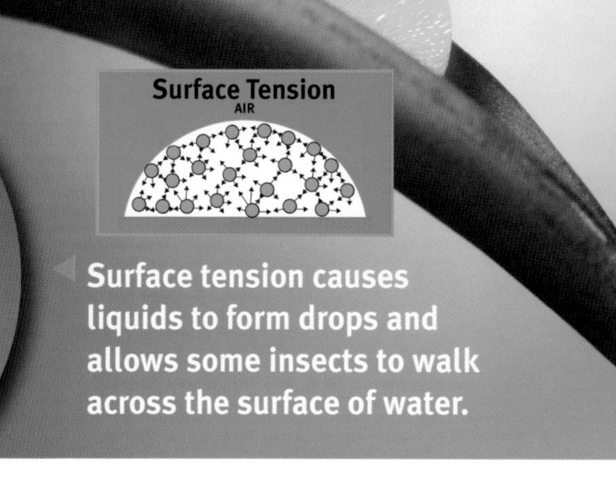

Surface tension causes liquids to form drops and allows some insects to walk across the surface of water.

Viscosity

Some liquids flow fast. Other liquids flow slowly. Viscosity (vis-KAH-sih-tee) is the resistance of a liquid to flowing. A liquid with a high viscosity flows slowly. A liquid with a low viscosity flows fast. Honey has a high viscosity. Vinegar has a low viscosity. Think of viscosity as how thick a liquid is.

Viscosity usually decreases as a liquid gets warmer. If you heat molasses, it will become thin and easy to pour. If you cool cooking oil in the refrigerator, it will become thick and difficult to pour.

Everyday Science

Bubbles

You can't make a bubble with ordinary tap water, but you can with soapy water. The reason is that soap decreases the surface tension of water. Soap molecules are long chains of atoms. When soap is added to water, one end of the chain attracts water molecules at the surface. This increases the distance between the water molecules. The farther apart they are, the weaker the pull between them is. As a result, the surface tension decreases and a bubble is able to form.

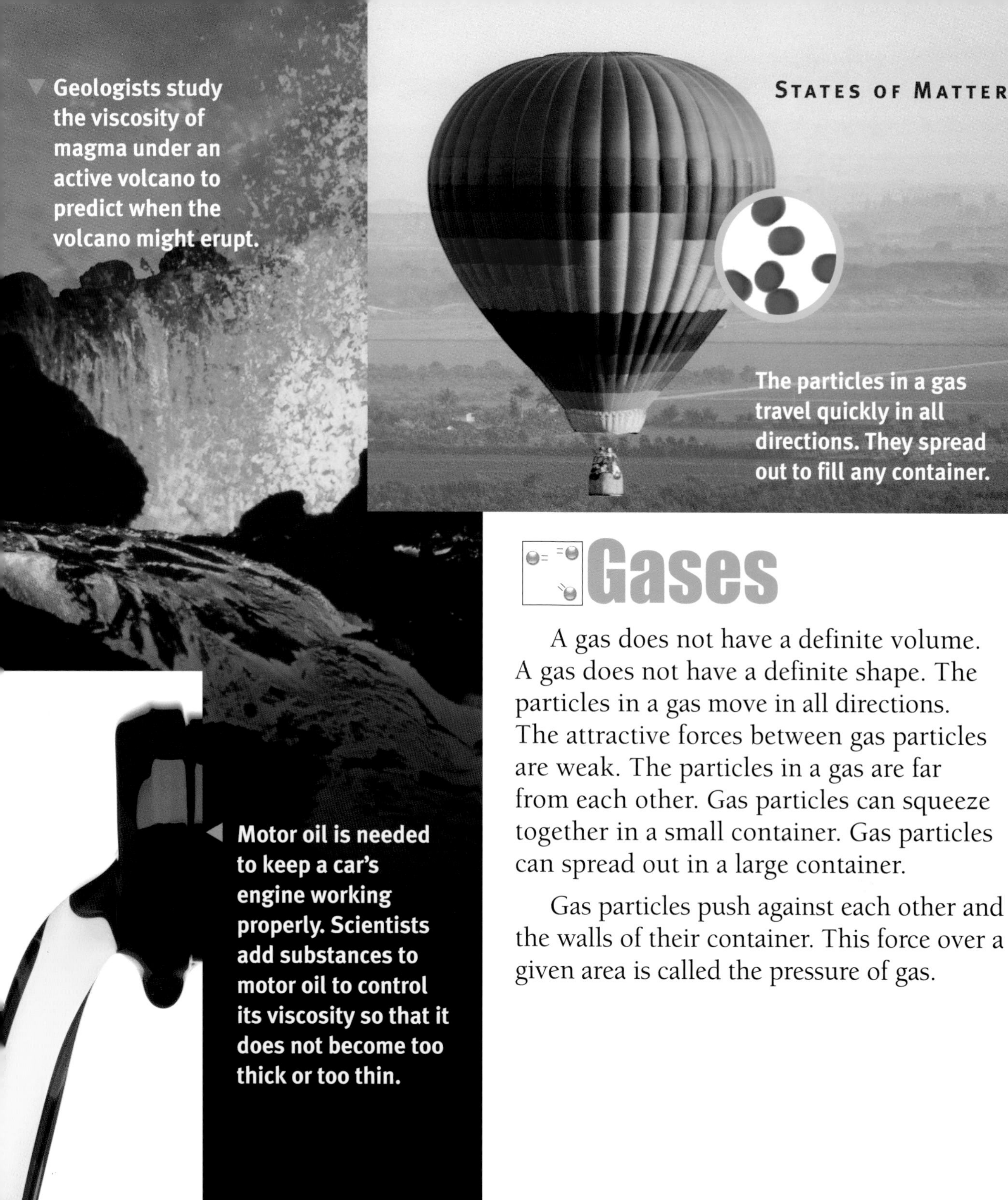

Geologists study the viscosity of magma under an active volcano to predict when the volcano might erupt.

The particles in a gas travel quickly in all directions. They spread out to fill any container.

Motor oil is needed to keep a car's engine working properly. Scientists add substances to motor oil to control its viscosity so that it does not become too thick or too thin.

Gases

A gas does not have a definite volume. A gas does not have a definite shape. The particles in a gas move in all directions. The attractive forces between gas particles are weak. The particles in a gas are far from each other. Gas particles can squeeze together in a small container. Gas particles can spread out in a large container.

Gas particles push against each other and the walls of their container. This force over a given area is called the pressure of gas.

As more air is pumped into the globe, there are more particles to collide. The result is collisions that occur more often and more intensely. This causes the pressure to increase.

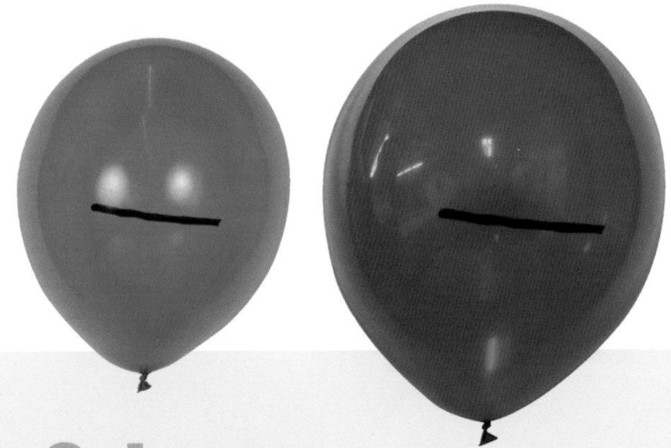

Hands-On Science

Changing the Temperature of a Gas

Time:	One hour
Materials:	Balloon, marker, tape measure, freezer
Step 1:	Blow up a balloon with air.
Step 2:	Use the marker to draw a line on the balloon.
Step 3:	Measure and record the length of the line you drew.
Step 4:	Place the balloon in the freezer for one hour.
Step 5:	Remove the balloon from the freezer. Measure the length of the line you drew. Explain your observations.

Everyday Science

Popcorn

Kernels of popcorn contain water. When a kernel is heated, the water molecules start moving faster. They collide with one another and the walls of the kernel more often. This causes an increase in the pressure inside each kernel. Eventually, the pressure becomes so great that the kernel pops. The result is delicious popcorn you can enjoy!

Plasmas

The fourth state of matter is plasma. Plasma exists at very high temperatures. Atoms are made up of even smaller particles. One of those particles is the electron. The atoms of a hot gas move fast. If the atoms of a hot gas hit one another, some electrons can get knocked out of the atoms. The result is a plasma. Plasma has no definite shape or volume. Plasmas are very rare.

Natural plasmas are rare on Earth. This is because Earth is too cold for matter to reach this state. People can use electricity to make plasmas. We can find plasmas in fluorescent lightbulbs and neon signs.

Scientists think that about ninety-nine percent of the universe is in the plasma state. Stars are a source of plasma.

▲ Like other stars, the sun provides the conditions necessary for matter to exist in the plasma state.

Science to Science

LIFE SCIENCE

The name for the fourth state of matter, plasma, came from the name of blood plasma. In 1927, scientist Irving Langmuir chose this name because he felt that matter in the plasma state behaved like plasma in the blood.

Changes in State

The state of matter depends on energy. In science, **energy** (EH-ner-jee) is the ability to do work or cause change. When matter is heated, it absorbs energy. When matter cools, it loses energy. As particles gain energy, they move faster and farther apart. As particles lose energy, they move more slowly and stay closer together. Matter can change its state by gaining or losing energy.

Melting

The change in state from a solid to a liquid is called melting. As a solid is heated, it absorbs energy. This causes the particles to vibrate faster. Particles can break free from their fixed positions. Particles then enter the liquid state. Most pure substances melt at a specific temperature. This temperature is called the melting point. Ice melts at 0°C (32°F).

Melting involves a change from the solid state to the liquid state. Metals can be heated until they become liquids that can be poured into molds.

Freezing

The change in state from a liquid to a solid is called freezing. This change is the reverse of melting. As a liquid loses energy, its particles slow down. The particles become locked in place and form a solid. The temperature at which freezing occurs is called the freezing point. The freezing point is the same as the melting point. The freezing point of water is also 0°C (32°F).

Most substances get more dense as they freeze. This is because the particles become packed more tightly together. Water does not behave this way. Solid ice is less dense than liquid water. Ice can float in water because ice is not as dense as water.

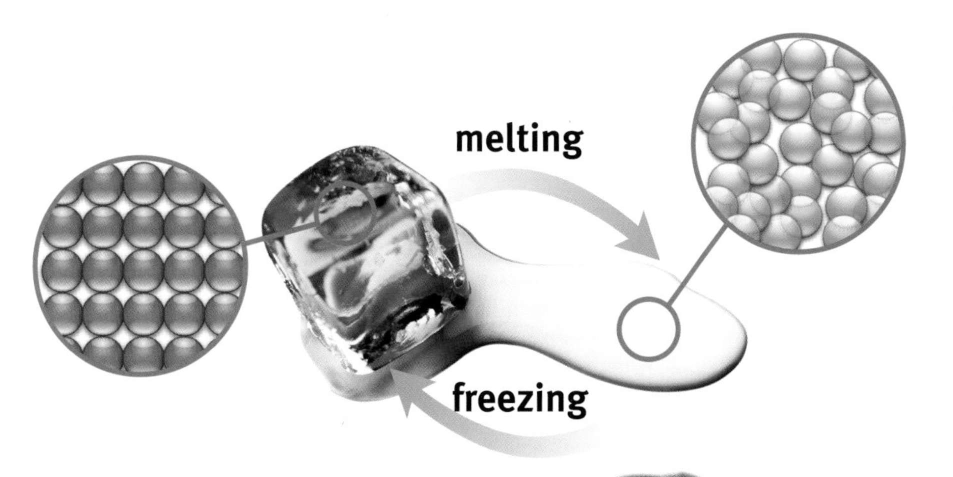

◄ A solid melts when its particles absorb enough energy to start moving past one another. A liquid freezes when it loses so much energy that its particles become locked in their positions.

✔ CHECKPOINT

Read More About It

Rocks melt into magma as part of the rock cycle. Read about magma in your school library or local library. Find out how magma forms below Earth's surface and erupts from volcanoes as lava.

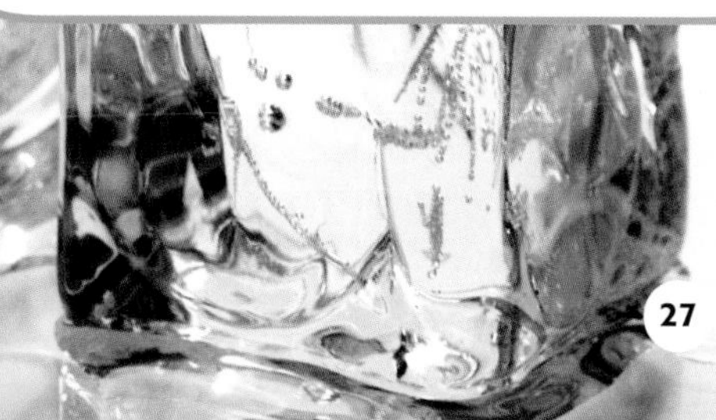

Hands-On Science

How does salt affect the freezing point of water?

TIME: Overnight

MATERIALS: Two plastic cups, water, salt, spoon, marker

SAFETY: Have paper towels on hand to clean up any spills.

STEP 1: Fill both cups about three-quarters full with water. Make sure to use the same amount of water at the same temperature.

STEP 2: Add four heaping spoonfuls of salt to one of the cups. Use the marker to write "Salt" on the outside of this cup.

STEP 3: Place both cups in the freezer. Note the time.

STEP 4: Check the cups every hour for five hours. Then leave the cups overnight and check them again in the morning.

STEP 5: What happened to the water in each cup? Did the plain water freeze? If so, how long did it take? Did the water with salt freeze? If not, how did the water appear in the morning?

What is the effect on the freezing point of water by adding salt to water?

Vaporization

The change in state from a liquid, or a solid, to a gas is called vaporization (vay-per-ih-ZAY-shun). This change occurs when the particles gain enough energy to enter the gas state. One type of vaporization is evaporation (ih-va-puh-RAY-shun). Another type of vaporization is boiling.

Vaporization at the surface of a liquid is called evaporation. Some particles have enough energy to escape the liquid. These particles become a gas. A puddle on the sidewalk disappears because the water evaporates.

Sometimes, a liquid changes to a gas below the surface. This is called boiling. Bubbles form in the liquid and rise to the surface. The temperature at which a liquid boils is called the boiling point. The boiling point of water is 100°C (212°F).

▲ Liquids can change to gases through evaporation and boiling. Evaporation can occur at temperatures below the boiling point. Water that enters the gas state through evaporation is called water vapor. Water that enters the gas state through boiling is called steam.

Career

CANDYMAKER

Making candy can be fun and delicious. It is also a precise science. A career as a candymaker means being both an artist and a chemist. One reason is that heating sugar involves knowing about solutions, temperature, and vaporization. Candymakers can work with large manufacturers or in individual stores.

Condensation

The change in state from a gas to a liquid is called condensation (kahn-den-SAY-shun). This change is the reverse of vaporization. Particles of a gas can lose energy. Particles of a gas can slow down and form a liquid. The fog you see on your bathroom mirror after a hot shower is condensation.

Look at the geyser in the photo. Do you see steam? If you answered "No," you are right. You cannot see water in the gas state. This means you cannot see water vapor or steam. The white cloud over the geyser is not a gas. Those clouds form when steam condenses into drops of water.

You cannot see water in the gas state. A cloud such as this one forms when steam condenses into drops of liquid water.

Science to Science

EARTH SCIENCE & PHYSICAL SCIENCE

In nature, water changes from one state of matter to another in the water cycle. When water on Earth's surface absorbs enough energy, it evaporates into water vapor in the air. When that water vapor cools, it condenses into drops of liquid water that form clouds. The drops that become heavy enough fall to the ground as precipitation, usually in the form of rain. If the drops lose more energy, they freeze into snow or ice. Then the process continues in a never-ending cycle.

The Root of the Meaning:
THE WORD
SUBLIMATION
comes from the Latin *sublimis*,
which means "lifted up."

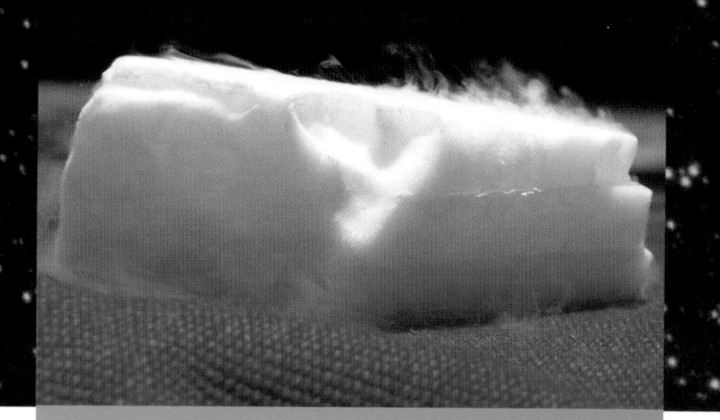

Dry ice sublimes into a gas. In a similar way, the frozen gases in the tail of a comet sublime as a comet approaches the sun. The tail is what makes it possible to see the comet from Earth.

Sublimation

A solid can change directly into a gas without passing through the liquid state. This process is called sublimation (suh-blih-MAY-shun). Dry ice is the solid form of carbon dioxide (CO_2). Dry ice sublimates to form a gas. A block of dry ice quickly shrinks and makes a cloud. Dry ice disappears as a gas without leaving a messy liquid behind.

The gas you see around dry ice is a cloud of liquid water drops that form in the air through condensation. The dry ice absorbs heat energy from the air around it. The dry ice makes the air colder. The colder air cools water vapor until it turns into a liquid.

Graphing Changes of State

We can show changes of state on a graph. The graph shows water in the solid state, as ice. When water absorbs energy, its temperature goes up to water's melting point. At its melting point, the water continues to absorb energy. But the water's temperature does not change. Then solid ice changes into liquid water.

Once all the ice changes into liquid water, the temperature goes up again. The temperature of the liquid rises until the liquid reaches the boiling point—100°C (212°F). Then the liquid water changes into water vapor. Once the change from the liquid state to the gas state is complete, the temperature goes up again.

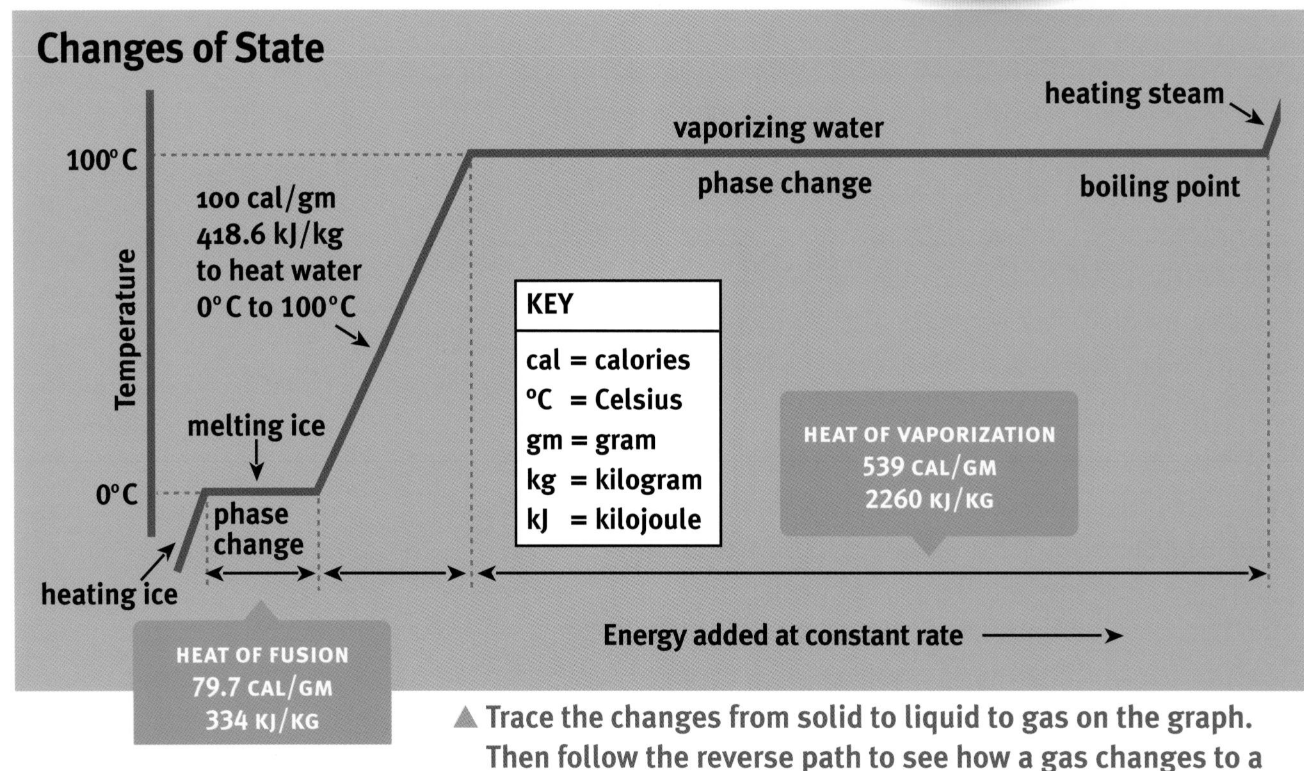

▲ Trace the changes from solid to liquid to gas on the graph. Then follow the reverse path to see how a gas changes to a liquid and then to a solid as it loses energy.

Summing Up

- Matter can exist as a solid, a liquid, a gas, or a plasma.
- The particles of a solid are held in fixed positions and can only vibrate.
- The particles of a liquid are packed together like those of a solid, but they are free to move past one another.
- In a gas, the particles are held only loosely together. They move quickly in all directions.
- If gases are heated to high temperatures, electrons are knocked off atoms and plasma forms.
- Matter can change from one state to another by gaining or losing energy.
- During melting, matter absorbs energy and changes from a solid to a liquid.
- During freezing, a liquid loses energy and changes to a solid.
- During vaporization, matter absorbs energy and changes from a liquid to a gas. Vaporization can occur as evaporation or boiling.
- The reverse of vaporization is condensation, during which a gas loses energy and changes to a liquid. Sublimation is the change directly from a solid to a gas.

Putting It All Together

Choose one of the research activities below. Work independently, in pairs, or in small groups. Share your responses and listen to other groups present their responses.

1. Along with other groups of students in your class, act out each of the states of matter. Your actions should relate the arrangement and motion of particles to the state of matter.

2. Make a table to summarize the changes of state. Identify changes between solids and liquids, liquids and gases, and solids and gases. Include examples of each.

3. The graph on page 32 relates heat energy to the state of matter. Explain this relationship.

Properties of Matter

What are the properties of matter and how can these properties change?

You have read about some of the properties of matter. Matter has two types of properties. We can use physical properties to describe matter. We can also use chemical properties to describe matter.

Physical Properties

A **physical property** (FIH-zih-kul PRAH-per-tee) is a trait you can see without changing the matter. Color, smell, texture, and density are physical properties. Melting point, boiling point, and the state of matter are also physical properties. Look at the chart on page 35 to see more.

Some physical properties are known as extensive properties. These depend on the amount of matter that you have. Length, mass, volume, and weight are examples of extensive properties. Extensive properties change as the amount of matter changes. Other physical properties do not depend on the amount of matter. These properties are called intensive properties. Intensive properties stay the same no matter how much matter you have. The state of matter, boiling point, melting point, and density are other examples of intensive properties.

Essential Vocabulary

chemical change page 38
chemical property page 38
chemical reaction page 38
law of conservation of mass page 40
physical change page 36
physical property page 34

Physical Properties of Matter

PROPERTY	EXPLANATION
State	Form or phase of the matter (solid, liquid, gas, or plasma)
Boiling point	Temperature at which the matter boils
Melting point	Temperature at which the matter melts
Color	How the matter reflects light from the visible spectrum
Odor	How the matter smells
Texture	How the matter feels
Density	Mass per unit volume of the matter
Malleability	Ability to be pounded into thin sheets
Ductility	Ability to be pulled into the shape of a wire
Luster	How the matter reflects light
Conductivity	Ability to let heat or electricity pass through
Length	Measure of how long the matter is
Mass	Measure of how much matter is in an object
Volume	Measure of how much space an object takes up

▲ Scientists use physical properties to distinguish one sample of matter from another.

Physical Changes

Origami is the art of folding paper. When you do origami, you make a physical change in matter. A **physical change** (FIH-zih-kul CHANJE) alters the form or appearance of matter. A physical change does not change the matter's makeup. Fold a piece of paper. The paper is still paper even though it looks different. You can undo physical changes. You can unfold a folded piece of paper.

A change of state is a physical change. Matter can change from one state to another, but its makeup stays the same. Think about what happens when an ice cube melts. Both the ice cube and the liquid water are made of water molecules. Only the order of the molecules is different. The pictures show more examples of physical changes.

▼ When these sheets of paper were folded into shapes, no new substances were formed. Folding paper is an example of a physical change in matter.

Cutting and sanding wood or shaping are physical changes. All of the physical changes shown have something in common—they involve a change in the form of matter, but not its composition.

CHECKPOINT

VISUALIZE IT

Make a table in your science notebook. Label one column "Physical Change" and another column "Example." Write the type of change you observe and a brief description of the example. Share your table with your classmates.

Chemical Properties

A **chemical property** (KEH-mih-kul PRAH-per-tee) of matter is an ability to change into different substances. You must change matter to see a chemical property. For example, coal can burn. This is a chemical property called flammability (fla-muh-BIH-lih-tee). You cannot see this property by looking at coal. You have to burn the coal to know that coal has this property. Coal turns into a different substance when it burns. Some matter can rust. This is another chemical property.

Chemical Changes

A chemical change happens when coal burns or iron rusts. A **chemical change** (KEH-mih-kul CHANJE) happens when one type of matter changes into a different type of matter. A chemical change makes new substances. These new substances have different properties from the original substances. A chemical change is the result of a **chemical reaction** (KEH-mih-kul ree-AK-shun).

▼ During all chemical changes, new substances are formed.

▲ Coal has the ability to burn under the right conditions. This property of coal cannot be identified by simply looking at the coal. The ability to burn is a chemical property.

The substances that are changed in a chemical reaction are called the reactants. The substances that are formed are called the products. During a chemical reaction, reactants change into products. The products have different properties than the reactants. A chemical change cannot be physically undone.

▲ When silver tarnishes, reactants change into products with different physical and chemical properties.

Think about what happens when a silver plate tarnishes. The silver reacts with sulfur in the air. Together the silver and sulfur form a black material. Silver and sulfur are reactants in this reaction. The black material is the product. The properties of the black material are different from those of silver or sulfur. We can describe this reaction like this:

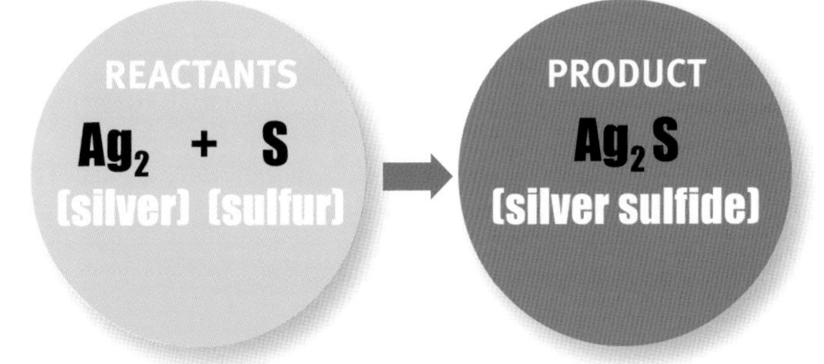

Science to Science

LIFE SCIENCE & PHYSICAL SCIENCE

Every autumn, trees in parts of the Northern Hemisphere turn spectacular colors as a result of chemical changes. In spring and summer, tree leaves are green because of a pigment called chlorophyll. Leaves also contain pigments that appear as shades of yellow and orange, but they are hidden by the chlorophyll. In autumn, changes in weather and the length of daylight cause trees to stop producing chlorophyll. As a result, the green color fades and the other colors can be seen.

They Made a Difference

Antoine Lavoisier (1743–1794)

Unlike most scientists who came before him, Antoine Laurent Lavoisier conducted hundreds of controlled experiments to test his ideas. In so doing, he disproved many accepted ideas of the time. He identified several chemical elements, he proposed the law of conservation of mass, he recognized that substances burn in oxygen, and he played a major role in the development of the metric system. For his contributions, he will forever be remembered as the Father of Modern Chemistry.

Conservation of Mass

A French chemist named Antoine Lavoisier (ahn-TWAHN luv-WAH-zee-ay) did experiments with chemical reactions. Lavoisier proposed the **law of conservation of mass** (LAW UV kahn-ser-VAY-shun UV MAS), which says that matter is neither created nor destroyed during a chemical reaction. The mass of the reactants is equal to the mass of the products.

During a chemical reaction, the bonds between the atoms of the reactants break. The atoms then rearrange and new bonds form in the products. All of the atoms of the reactants are in the products. All of the atoms of the reactants are now arranged in a different way.

Below, we can see the chemical reaction that happens when water forms. Look at the atoms of each element. The reactants have four hydrogen atoms and two oxygen atoms. Are all of those atoms in the products? Yes, they are.

$$2H_2 + O_2 \rightarrow 2H_2O$$

▲ The reactants and the products each have four hydrogen atoms and two oxygen atoms. However, the atoms are arranged in a different way. Because the numbers of atoms have not changed, the mass is the same before and after the reaction.

Summing Up

- Matter can be described by physical properties and chemical properties.
- A physical property is one that can be observed without changing the composition of the matter.
- A chemical property is an ability to undergo change. Matter can undergo both physical and chemical changes.
- During a physical change, only the form or appearance of matter changes.
- During a chemical change, new substances with different properties are produced.
- In a chemical reaction, mass is conserved. This means that matter is neither created nor destroyed. Therefore, the mass of the reactants is always equal to the mass of the products.

Putting It All Together

Choose one of the research activities below. Work independently, in pairs, or in a small group. Share your responses with the class. Listen to other groups present their responses.

1. Write down several physical properties for a sheet of construction paper. Show how to cause the paper to undergo at least three different physical changes.

2. Create a journal with two columns on each page. For two days, write down any physical and chemical changes that you observe. Be sure to include descriptions of the physical and chemical properties of the examples before and after the change.

3. The law of conservation of mass described on page 40 explains that atoms are rearranged during a chemical change. They can neither be created nor destroyed. Create a model to show how this is true.

A Matter of Properties

You live in a world filled with matter. Matter can be a pure substance. Matter can be an element or compound. Matter can be a mixture of substances. Matter can be found as a solid, liquid, gas, or plasma. Each state of matter has different physical properties. Matter can change from one state to another. These are physical changes.

Matter can also experience chemical changes. We see chemical changes all the time. Silver tarnishes. A wagon wheel rusts. A flame burns on a stove. These are all chemical changes. In a chemical change, atoms of one type of matter react with other atoms to make a new substance. The properties of this new substance will differ from the properties of the original atoms. We can use these properties to describe matter.

How to Write an Observation Log

Every day you make many observations. You may observe what the weather conditions are, how long an activity takes, or the temperature of your food. When you make an observation, you collect information using your five senses: tasting, touching, seeing, hearing, and smelling. In science, an observation is both the process of collecting data and the data collected.

Once you make scientific observations, you need to keep track of them in an observation log. When you create an observation log, you keep a journal in which you record your observations in an orderly fashion. Someone looking at the log should be able to tell when you made the observation, what you observed, and any other information related to the observation. The information might be presented in a list or a chart. It might also include drawings or measurements.

One important thing to remember when keeping an observation log is that you should record only what you truly observe. Do not include information that you expected to observe but did not. Do not add observations that you missed or were unable to make.

Scientists keep observation logs until they have enough information to successfully complete their investigation. If you are observing a chemical change in a laboratory, your log might last for a few hours. If you are observing seasonal changes, your log might last for a year. For some topics, an observation log might last several years.

Try your hand at keeping an observation log. Choose a topic for which you will make observations. Think about the information you should include. Then create an observation log similar to the one on page 45.

Date	Time	Height
April 12	9:00 A.M.	12.0 cm
April 15	12:00 P.M.	13.5 cm
April 20	11:00 A.M.	16.2 cm

▲ The entries above are from the observation log of a researcher studying the growth of tomato plants.

Date	High Temperature	Low Temperature	Precipitation	Clear or Cloudy

▼ A researcher studying the moon might include such information as when the observation was made, the location from which the observation was made, and a drawing of the observation.

Date	Location	Conditions	Observation	Description
June 1	Front porch	Clear		1st quarter moon
June 8	Front porch	Cloudy		Waxing gibbous moon
June 15	Front porch	Rainy	N/A	Unable to see
June 16	Front porch	Partly cloudy		Full moon
June 24	Front porch	Clear		Waning gibbous moon

Glossary

atom	(A-tum) *noun* the smallest particle of an element that has the characteristics of that element; the basic building block of matter, consisting of protons, neutrons, and electrons (page 10)
chemical change	(KEH-mih-kul CHANJE) *noun* a change in which matter becomes a different type of matter with different chemical properties (page 38)
chemical formula	(KEH-mih-kul FOR-myuh-luh) *noun* a notation that shows the ratio of atoms of the combining elements in a compound (page 12)
chemical property	(KEH-mih-kul PRAH-per-tee) *noun* the ability of matter to change into a certain substance (page 38)
chemical reaction	(KEH-mih-kul ree-AK-shun) *noun* the process by which matter changes into a different type of matter with different chemical properties (page 38)
chemical symbol	(KEH-mih-kul SIM-bul) *noun* an abbreviation for a chemical element that consists of one or two letters (page 8)
compound	(KAHM-pownd) *noun* a pure substance that is formed when the atoms of two or more elements are held together by chemical bonds (page 10)
element	(EH-leh-ment) *noun* a pure substance that cannot be broken down into simpler substances (page 7)
energy	(EH-ner-jee) *noun* the ability to do work or cause change (page 26)

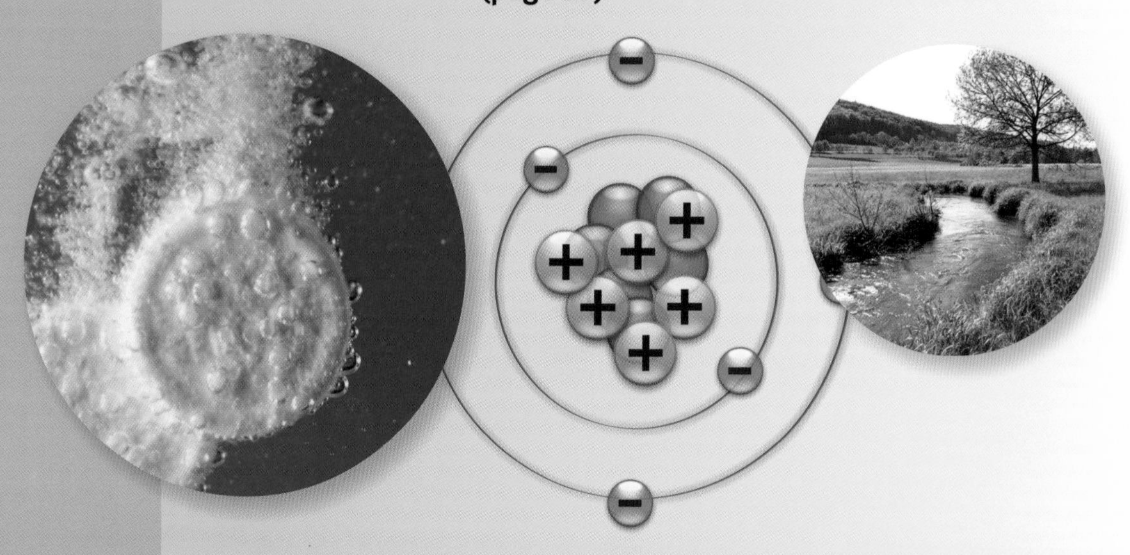

gas	(GAS) *noun* the state of matter that does not have a definite volume or a definite shape because the particles have so much energy that they travel quickly in all directions (page 19)
law of conservation of mass	(LAW UV kahn-ser-VAY-shun UV MAS) *noun* statement that says that matter is neither created nor destroyed during a chemical reaction (page 40)
liquid	(LIH-kwid) *noun* the state of matter that has a definite volume but not a definite shape because the particles have enough energy to flow past one another (page 19)
mass	(MAS) *noun* the amount of matter in an object (page 6)
matter	(MA-ter) *noun* anything that has mass and volume (page 6)
mixture	(MIKS-cher) *noun* a blend of two or more substances that are not chemically combined (page 13)
molecule	(MAH-leh-kyool) *noun* the smallest unit of an element or compound that keeps all of the characteristics of that element or compound (page 11)
physical change	(FIH-zih-kul CHANJE) *noun* a change in the form or appearance of matter (page 36)
physical property	(FIH-zih-kul PRAH-per-tee) *noun* a characteristic that can be observed that does not change the composition of the matter (page 34)
plasma	(PLAZ-muh) *noun* the state of matter that exists at high temperatures in which electrons are knocked out of atoms of a gas (page 19)
pure substance	(PYER SUB-stans) *noun* matter, including elements and compounds, that has a unique set of characteristics as compared with mixtures (page 13)
solid	(SAH-lid) *noun* the state of matter that has a definite shape and a definite volume because the particles are held in relatively fixed positions (page 19)
volume	(VAHL-yoom) *noun* the amount of space an object takes up (page 6)

Index